Prajakta Wasekar, U.M. Gokhale

High Performance Carry Select Adder Using Binary Excess Converter

GRIN Publishing

Bibliographic information published by the German National Library:

The German National Library lists this publication in the National Bibliography; detailed bibliographic data are available on the Internet at http://dnb.dnb.de .

Imprint:

Copyright © 2012 GRIN Verlag GmbH
Print and binding: Books on Demand GmbH, Norderstedt Germany
ISBN: 978-3-656-88584-9

This book at GRIN:

http://www.grin.com/en/e-book/288145/high-performance-carry-select-adder-using-binary-excess-converter

GRIN - Your knowledge has value

Since its foundation in 1998, GRIN has specialized in publishing academic texts by students, college teachers and other academics as e-book and printed book. The website www.grin.com is an ideal platform for presenting term papers, final papers, scientific essays, dissertations and specialist books.

Visit us on the internet:

http://www.grin.com/

http://www.facebook.com/grincom

http://www.twitter.com/grin_com

High Performance Carry Select Adder Using BEC

Prajakta S. Wasekar
G.H.Raisoni Institute of
Engineering & Technology for
Women, Nagpur

Prof.U.M.Gokhale
Professor, ETC Dept.
G.H.Raisoni Institute of
Engineering & Technology for
Women, Nagpur

Abstract: **Adders are one of the widely used digital components in digital integrated circuit design. Addition is the basic operation used in almost all computational systems. Therefore, the efficient implementation and design of arithmetic units requires the binary adder structures to be implemented in an equally efficient manner. A ripple carry adder has smaller area but less speed. A carry look-ahead adder is faster though its area requirements are high. Carry select adders (CSLA) lie in middle. In this work a novel carry select adder using Binary Excess Converter (BEC) is proposed. It provides good compromise between cost and performance thereby establishing a proper trade-off between time and area complexities. In this work Tanner EDA is used for the comparison of all adders – Ripple carry adder, Bitwise carry select adder, Square root carry select adder, proposed carry select adder using BEC.**

Keywords: **Carry Select Adder, Binary Excess Converter, Fast Adder.**

I. Introduction

In recent years, the increasing demand for high-speed arithmetic units in microprocessors, image processing units and DSP chips has paved the path for development of high-speed adders as addition is an indispensable operation in almost every arithmetic unit; also it acts as the basic building block for synthesis of all other arithmetic computations. To increase portability of systems and battery life, area and power are the critical factors of concern. Even in Servers and Personal Computers (PC), power dissipation is an important design parameter. In today's scenario, design of area-efficient and power-efficient high-speed logic systems is the one of the crucial areas of research in VLSI design. In digital adders, the speed of addition is limited by the time required by the carry to propagate through it. Depending on the area, delay and power consumption requirements, several adder implementations have been proposed. Ripple Carry Adders with the most compact design among all types of adders are slowest in speed.

Carry Select Adder (CSLA) is one of the fastest adders used in many data-processing processors to perform fast arithmetic functions. By gate level modification of CSLA architecture, we can reduce area and power.

The basic motive of this work is to design and develop an efficient less area and low power adder. From the structure of the CSLA, it is clear that there is scope for reducing the area and power consumption in the CSLA. This work uses simple and efficient gate-level modification to significantly reduce the area and power of the CSLA. The proposed design reduces area and power as compared with the regular SQRT CSLA with only a slight increase in the delay.

II. Regular Carry Select Adder

In regular carry select adder (CSLA) the optimization of power and area is possible trying different options for the logic style, differential cascode voltage switch, complementary pass-transistor logic, double pass-transistor logic, and swing restored CPL, and hybrid styles. The main advantage of CSLA [1] is to reduced propagation delay characteristics. This is realized by the use of parallel stages that results from multiple pairs of ripple carry adder (RCA). In ripple carry adder, carry-out of one stage is connected carry-in of next stage. The sum and carry out bits of any stage cannot be produced, until sometime after the carry–in of the stage occurs. This is due to the propagation delay in logic circuitry, which leads to a time delay in the addition process.

Therefore, for entire process in ripple carry adder the speed of addition is limited by the time required to propagate a carry through adder. The CSLA is use in many computational systems to alleviate the problem of carry propagation delay by independently generating multiple

1

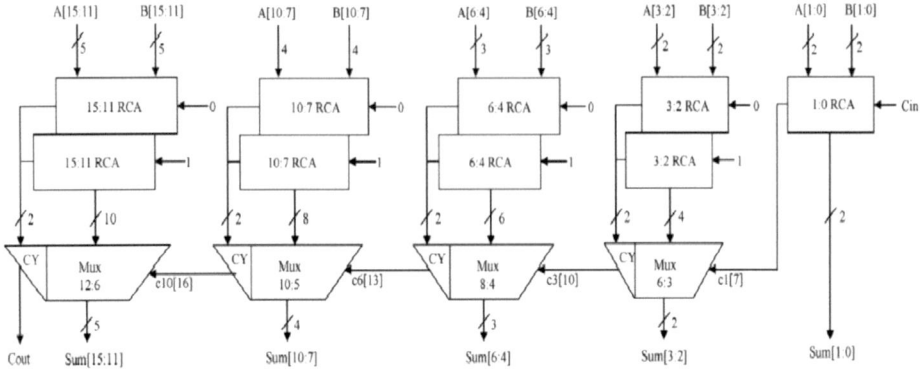

A[15:11] B[15:11] A[10:7] B[10:7] A[6:4] B[6:4] A[3:2] B[3:2] A[1:0] B[1:0]

Carries from CSLA and then select a carry to generate the sum.

III. Binary Excess Convertor (BEC)

The main idea of this work is to use BEC instead of the RCA with Cin=1 in order to reduce the area and power consumption of the regular CSLA. Fig 2 shows a 4bit BEC logic diagram consisting of XOR, AND gates and an inverter. Fig.3 a) shows the schematic view of 4-bit BEC. The main advantage of this BEC logic comes from the lesser number of logic gates than the n-bit Full Adder (FA) structure.

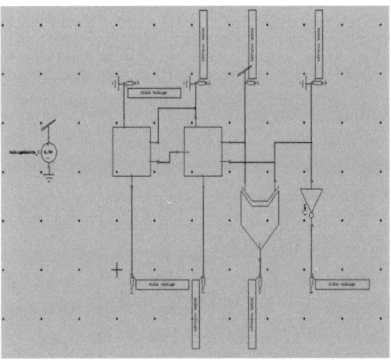

Fig.3 (a): Schematic view of 4-bit BEC

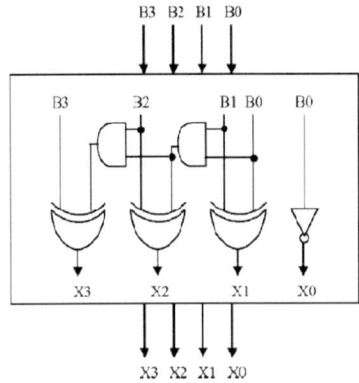

B3 B2 B1 B0

Fig 2: 4-bit Binary Excess Converter (BEC)

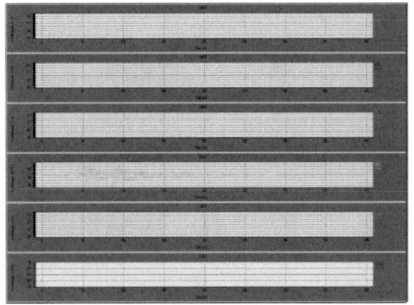

Fig. 3(b): Waveform of 4-bit BEC

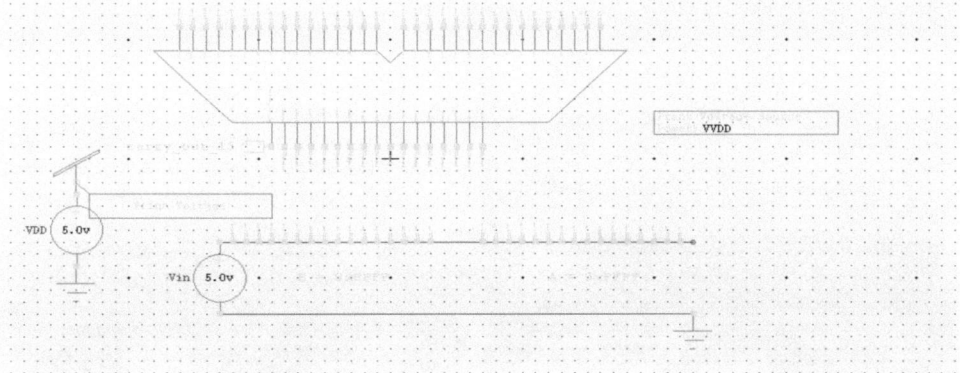

Fig.4: Schematic representation of modified CSLA

IV. Proposed Carry Select Adder

The proposed carry select adder (CSLA) is as shown in figure (4). It consists of RCA, BEC and multiplexer. Based on this modification 8-, 16--b square-root CSLA (SQRT CSLA) architecture has been developed and compared with the regular SQRT CSLA [1] architecture. The proposed design has to reduced area and power as compared with the regular SQRT CSLA with only a slight increase in the delay. Figure (2) shows the architecture of proposed BEC 2248 Efficient carry select adder. It shows a (BEC) Efficient Novel CSLA [1] that replaces the RCA with input carry as '1' block in conventional square root CSLA. This produces the two possible partial results in parallel and the multiplexer is use to select either the BEC output or the direct input according to the control signal 'Cin' The LSB's are added using conventional RCA. Once all the sums and carries are calculated, the final sums are computed using multiplexers having minimal delay. The multiplexer block receives the two sets of input and selects the final sum based on the select input from the previous stage. By using this modification area, power dissipation can be reduced. As shown in fig (1) RCA that is used in regular 16-bit SQRT CLSA is replaced by Binary Excess Convertor (BEC).

Thus, the number of gates will be reduced as compared to the regular SQRT CSLA. This can be helpful for reducing area,   delay as well as power consumption. The 4-bit BEC structure is as shown in fig (2).

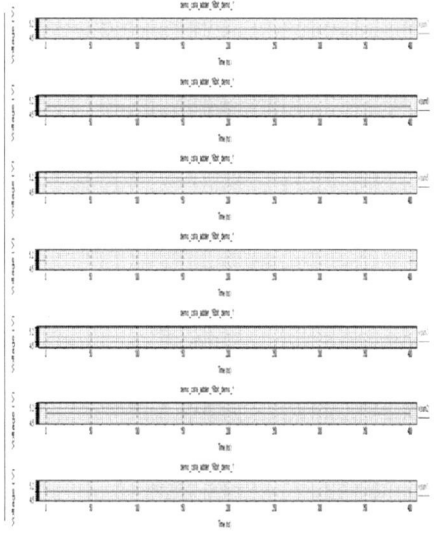

3

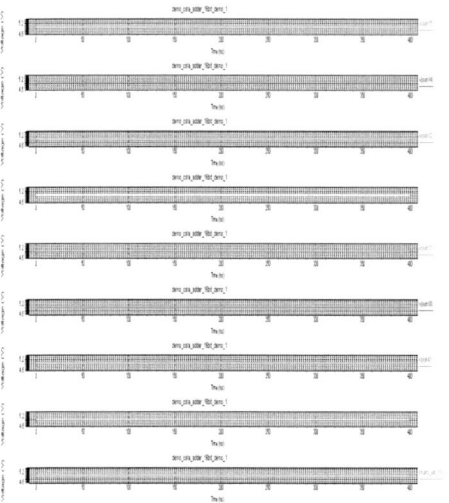

Fig.5: Simulation Waveform of modified CSLA

To reduce delay in CSLA certain modifications can be done in BEC in order to reduced area, delay and power consumption. Here we can reduce the size as well as number of gates by simply using BEC in place of RCA. The modified SQRT CSLA is then compare with regular SQRT CSLA in order to optimized power and area

V. Results

The design proposed in this work has been developed using Tanner EDA using typical libraries of TSMC 50nm technology. The synthesized netlist and their respective design constraints file are important imported to microwind encounter and are used to generate automated layout from standard cells and placement and routing.

Table below exhibits the simulation result of both regular and modified CSLA in the terms of area, delay and power. The area indicates the total cell area of the design in terms of number of MOSFETs.

Table 1: Comparison of 16-bit Regular & Proposed CSLA

Parameters	16-bit	
	Regular	Modified
Area (No. of MOSFET)	1771	1414
Power(uw)	464.9262	3.65
Current(nA)	92.985	73.06

VI. Conclusion

A simple approach is proposed in this paper is to reduce the area and power consumption of CSLA architecture. The reduced number of gates in this work offers the great advantage in the reduction of area and the total power. By comparing the proposed CSLA with regular CSLA, the area and power of the 16-b modified CSLA are significantly reduced by 15% and 20% respectively. The modified CSLA architecture is therefore, low area, low power, simple and efficient for VLSI hardware implementation. It would be interesting to test the design of the modified 128-b SQRT CSLA.

VII.   References

1. B. Ramkumar and Harish M Kittur, "Low-Power and *IEEE transactions* on very large scale integration (VLSI) systems vol. Area-Efficient Carry Select Adder".20, no. 2, February 2012.

2. Wonhak Hong, Rajashekhar Modugu, and Minsu Choi, Efficient Online Self-Checking Modulo Multiplier Design *IEEE* transactions on computers September 2011.

3. Mariano Aguirre-Hernandez, "CMOS Full-Adders for Energy-Efficient Arithmetic Applications" *IEEE transactions* on very large scale integration (VLSI) Systems June 2010.

4. B.Ramkumar, Harish M Kittur and P.Mahesh Kannan, "ASIC Implementation of Modified Faster Carry Save Adder", European Journal of Scientific Research, vol.42 no.1, pp.53-58, 2010.

5. Y. He, C. H. Chang, and J. Gu, "An area efficient 64-bit square root carry-select adder for low power applications," in *Proc. IEEE Int. Symp. Circuits Syst., 2005, vol. 4, pp. 4082–4085.*

6. J. M. Rabaey, *Digital Integrated Circuits—A Design Perspective.* Upper Saddle River, NJ: Prentice Hall, 2001.

7. R. Zimmerman and W. Fichtner, "Low-power logic styles: CMOS versus pass-transistor logic," *IEEE J.*

*Solid-State Circuits, vol. 32, no.*7, pp. 1079–1090, Jul. 1997.

8. N. Weste and K. Eshraghian, Principle of CMOS VLSI Design, A System Perspective. Reading, MA: Addison-Wesley, 1988, *ch.*5

9. O.J.Bedrij, "Carry-Select Adder", IRE Transactions on Electronic Computers, Pp. 340-344,1962.

10. V.G. Oklobdzija, "High-Speed VLSI Arithmetic Units: Adders and Multipliers", in "Design of High-Performance Microprocessor Circuits", Book edited by A.Chandrakasan, IEEE press, 2000.